SHARK SHOCK!

HAMMERHEAD SHARK

By Martha London

Consultant: Erin McCombs
Educator, Aquarium of the Pacific

BEARPORT
PUBLISHING

Minneapolis, Minnesota

Credits

Cover and title page, © EXTREME-PHOTOGRAPHER/iStockPhoto, © Yuri Samsonov/Shutterstock, © Elenamiv/Shutterstock; 3, © nabil refaat/Shutterstock; 4-5, © EXTREME-PHOTOGRAPHER/iStockPhoto; 6-7, © Alastair Pollock Photography/Getty; 8-9, © negaprion/iStockPhoto; 10-11, © Gerard Soury/Getty; 12, © Eric Isselee/Shutterstock; 12-13, © FtLaudGirl/iStockPhoto; 14-15, © Colors and shapes of underwater world/Getty; 16-17, © Jeff Rotman /Alamy; 18-19, © Marie-Elizabeth Mali/iStockPhoto; 20-21, © Ken Griffiths/Shutterstock; 22, © frantisekhojdys/Shutterstock; 22-23, © nabil refaat/Shutterstock; 24, © nabil refaat/Shutterstock.

President: Jen Jenson
Director of Product Development: Spencer Brinker
Senior Editor: Allison Juda
Associate Editor: Charly Haley
Designer: Colin O'Dea

Library of Congress Cataloging-in-Publication Data

Names: London, Martha, author.
Title: Hammerhead shark / by Martha London.
Description: Minneapolis, Minnesota : Bearport Publishing Company, [2022] | Series: Shark shock! | Includes bibliographical references and index.
Identifiers: LCCN 2021034176 (print) | LCCN 2021034177 (ebook) | ISBN 9781636915326 (library binding) | ISBN 9781636915418 (paperback) | ISBN 9781636915500 (ebook)
Subjects: LCSH: Hammerhead sharks--Juvenile literature.
Classification: LCC QL638.95.S7 L66 2022 (print) | LCC QL638.95.S7 (ebook) | DDC 597.3/4--dc23
LC record available at https://lccn.loc.gov/2021034176
LC ebook record available at https://lccn.loc.gov/2021034177

Copyright © 2022 Bearport Publishing Company. All rights reserved. No part of this publication may be reproduced in whole or in part, stored in any retrieval system, or transmitted in any form or by any means, electronic, mechanical, photocopying, recording, or otherwise, without written permission from the publisher.

For more information, write to Bearport Publishing, 5357 Penn Avenue South, Minneapolis, MN 55419. Printed in the United States of America.

CONTENTS

Ready for a Bite 4
Big and Small 6
Close to Shore 8
Hammer for a Head 10
Looking for Lunch 12
Picnic in the Sand 14
Danger from the Surface 16
Gathering Together 18
Pups in the Shallows 20

More about Hammerhead Sharks 22
Glossary 23
Index 24
Read More 24
Learn More Online 24
About the Author 24

Ready for a Bite

A hammerhead shark sneaks up behind a stingray. In a sudden burst of speed, the shark hits the ray with its wide head. Before the stingray has time to get away, the shark pins its **prey** to the sand. *Chomp!* Dinner is served.

Hammerhead sharks sometimes gather near sunken ships to hunt for food.

Big and Small

Hammerhead sharks are known for their wide, flat heads and fierce hunting skills. But not all of these **predators** are large. There are 10 **species** of hammerhead sharks that range greatly in size. The smallest are only about 3 feet (0.9 m) long—about as long as a bus seat. But the largest are usually about 13 ft (4 m). That's around the length of four bus seats side by side.

> Great hammerhead sharks are the largest hammerhead species. Scalloped bonnethead sharks are the smallest.

A great hammerhead shark

Close to Shore

Hammerheads—large and small—live in warm waters close to the shore. But hammerhead sharks are always on the move. They travel with the seasons. During the summer, some species swim to cooler areas north or south of the **equator**. Then, in the winter, the sharks swim closer to the equator, where the water is warmer.

Scientists tracked one hammerhead swimming about 390 miles (630 km) in just 2 weeks!

HAMMERHEAD SHARKS AROUND THE WORLD

Where hammerhead sharks live

Hammer for a Head

These fish slice through the water with wide, flat heads. The hammer-like shape gives the sharks their name. But each species has a slightly different shape. Some are curved at the front. Others lead the way with straighter heads. The heads of the smallest species aren't in the shape of a hammer at all—they're more like a shovel!

Some species of hammerheads are born with curved heads that straighten out as they get older.

Looking for Lunch

Eyes and nose holes on either end of hammerheads' wide heads help them see and smell everything around them—including lunch. Sharks also use special **sensors** in their heads to find their food. These sensors feel the **electrical signals** of other animals nearby.

Larger hammerheads hunt bigger prey, such as stingrays or large fish. The smaller sharks attack smaller prey, including crabs.

With eyes so far apart, a hammerhead shark can see all the way around itself with just a simple turn of the head.

Picnic in the Sand

When a large hammerhead finds prey, the shark hits the creature with its wide head. The blow **stuns** the prey animal. Then, the shark uses its sharp teeth to take a bite out of its meal. The hammerhead does not chew its food. Instead, it swallows each piece whole. *Gulp!*

A hammerhead's teeth look like tiny saws. They can cut through muscle and bone.

Danger from the Surface

Some hammerhead sharks become prey themselves. Small hammerheads are eaten by larger sharks. Larger adults have very few predators.

However, all sharks face danger from humans. Some people hunt hammerhead sharks. Because of this threat, many species of hammerheads are **endangered**.

Even fishers who do not hunt hammerheads may accidentally hurt the fish while using nets to catch other creatures.

Gathering Together

Some kinds of larger hammerheads come together into groups, called schools. They swim around with other sharks during the day. Then, they break up at night to hunt and travel. Others live alone and come together only every couple of years to **mate** before they head out on their own again.

A school of scalloped hammerhead sharks may have more than 200 sharks in it!

Pups in the Shallows

Hammerhead sharks have their babies in **shallow** water. Baby sharks, called pups, are on their own right away. They stay in the shallow waters until they grow big enough to take care of themselves deeper under the waves. When they are between 5 and 17 years old, they can have babies of their own.

Most species of hammerhead sharks live between 20 and 30 years in the wild. But some may live more than 40 years.

21

More about Hammerhead Sharks

The longest hammerhead ever recorded was 20 feet (6.1 m) long!

Hammerhead sharks' heads are soft when they are born and harden as they get older.

Hammerhead sharks tan. They get darker when they spend a lot of time close to the surface of the water.

One type of hammerhead shark eats both plants and animals.

Scientists think hammerheads may have been around for more than 50 million years.

When larger hammerhead sharks feast, they may rip pieces off larger prey by shaking their heads back and forth.

Glossary

electrical signals information in the form of weak electricity surrounding some creatures

endangered in danger of dying out completely

equator the imaginary line around the middle of Earth

mate to come together to have young

predators animals that hunt and eat other animals

prey animals that are hunted by other animals

sensors things used to find or feel a change in the surroundings

shallow not deep

species groups that animals are divided into, according to similar characteristics

stuns shocks something so much that it is unable to move

Index

equator 8
eyes 12–13
food 4, 12, 14
head 4, 6, 10–14, 22
hunt 4, 6, 12, 16–18
mating 18
predators 6, 16
prey 4, 12, 14, 16, 22
pups 20
schools 18–19
sensors 12
species 6–8, 10–11, 16, 21
teeth 14–15

Read More

Nixon, Madeline. *Hammerhead Shark (Sharks).* New York: AV2, 2019.

Twiddy, Robin. *Hammerhead Shark (Teeth to Tail).* New York: KidHaven Publishing, 2020.

Learn More Online

1. Go to **www.factsurfer.com** or scan the QR code below.
2. Enter "**Hammerhead Shark**" into the search box.
3. Click on the cover of this book to see a list of websites.

About the Author

Martha London lives in Minnesota with her cat. She's written more than 100 books for young readers!